BEANS AND PEAS

Susanna Miller

Illustrations by John Yates

Carolrhoda Books, Inc./Minneapolis

All words that appear in **bold** are
explained in the glossary on page 30.

First published in the U.S. in 1990 by
Carolrhoda Books, Inc.

Library of Congress Cataloging-in-Publication Data

Miller Susanna.
    Beans and peas / Susanna Miller.
       p.   cm. — (Foods we eat)
    Summary: Describes beans and peas, the history of their
cultivation and use, and their role in industry and diet. Includes
some recipes.
    ISBN 0-87614-428-8 (lib. bdg.)
    1. Legumes — Juvenile literature. 2. Legumes as food — Juvenile
literature. 3. Cookery (Legumes) — Juvenile literature. [1. Beans.
2. Peas.] I. Title. II. Series: Foods we eat (Minneapolis, Minn.)
SB177.L45M55   1990
635'.65 — dc20   89-28449
                                                                          CIP
                                                                           AC

Printed in Italy by G. Canale C.S.p.A., Turin
Bound in the United States of America

1 2 3 4 5 6 7 8 9 10 99 98 97 96 95 94 93 92 91 90

Contents

What are beans and peas?

Beans and peas are types of legumes. Legumes are plants that grow seeds within pods. Usually both the seeds and the pods are edible. Legumes are a useful source of food because they are inexpensive and highly **nutritious**. Some beans, such as broad and runner beans, are sold fresh. Many other beans are sold as dry beans because they are eaten after they have become fully ripe.

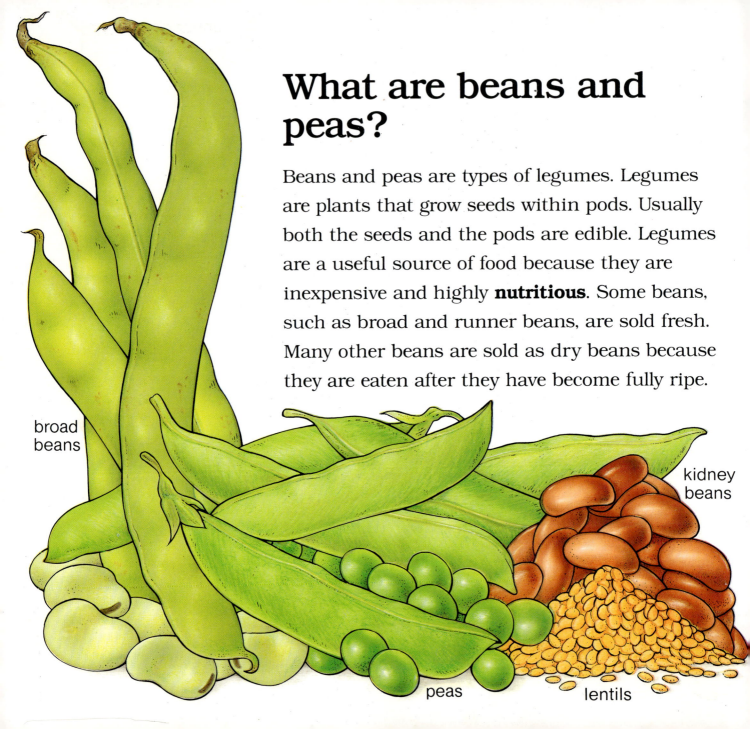

broad beans

kidney beans

peas

lentils

Dry beans take longer to cook than fresh beans, but they will keep for many months without going bad.

One type of bean nearly everyone has eaten is the kidney bean. Kidney beans are often combined with a tasty tomato sauce to make baked beans. Other popular types of beans include lima beans, lentils, mung beans, and soybeans. Soybeans contain nearly two times more **protein** than chicken and three times more protein than beef.

runner beans

soybeans

chickpeas

butter beans

Beans and peas in the past

Scientists have been able to prove that peas grew in Thailand as long ago as 9750 B.C. People began **cultivating** plants around 9000 B.C.

Legumes were among the very first crops to be farmed. The ancient Egyptians, Greeks, and Romans all cultivated various legumes. We know

A crop of soybeans drying on a Chinese farm. Soybeans were first grown in China 5,000 years ago.

that the ancient Hebrews ate beans and peas because legumes are mentioned several times in the Bible.

Legumes have been used in other ways than just as a food. The ancient Greeks and Romans used beans for casting their vote in elections. In ancient Egypt, beans were at one time considered a symbol of life, although at another time they were believed to contain the souls of the dead. In medieval times, beans were used in medicine and sorcery.

These Egyptian women are sifting lentils in preparation for a feast. Legumes have been cultivated in Egypt since ancient times.

7

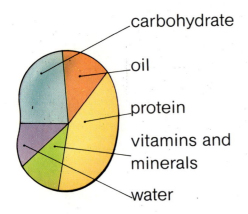

carbohydrate

oil

protein

vitamins and minerals

water

Above: This diagram shows the variety of nutrients in a soybean. Soybeans contain more protein than any other bean.

Right: A meal of beans on toast provides us with a combination of amino acids from legumes and grains.

The food value of beans and peas

Legumes are very good for a healthy diet. They are high in fiber and protein, and low in sugar and fat. Fiber helps the digestive system work properly, and protein builds new cells and repairs damaged tissues. Because they are low in sugar and fat, eating legumes will not increase our risk of heart disease.

Protein is made up of many **amino acids.** Our

bodies manufacture most amino acids, but there are eight essential amino acids we can only get by eating certain foods. All eight essential amino acids are not usually found in one type of food, so we need to combine foods.

Vitamins and minerals are found in small quantities in food. Our bodies need them only in tiny amounts, but without them we would suffer from diseases. Legumes contain important vitamins and minerals, such as vitamin B, iron, and phosphorus.

A market stall in Ankara, Turkey, displays legumes, dried fruit, and seeds for sale.

9

Farming beans and peas

Beans and peas belong to the family of plants with the scientific name *leguminosae.* There are 14,000 to 17,000 species of *leguminosae,* which grow nearly all over the world. But legumes grow best in warm, temperate climates.

The United States is one of the world's largest **exporters** of legumes. Great Britain buys beans from the United States for the 900 million cans of

Watering a crop of snowpeas in Thailand

baked beans eaten there every year. The world's largest producer of lentils is India. The United States, Argentina, Egypt, Ethiopia, Morocco, Spain, Syria, and Turkey also grow lentils.

Inspecting soybean seedlings on a farm in Brazil

At harvesttime, the legume plants are cut and left in the fields to dry. Then mechanical harvesters gather up the plants and pass them through threshing cylinders, which separate the pods from the waste part of the plant. In very dry areas, plants may be left to dry out before being harvested.

From harvest to store

After the crops have been harvested, the pods are taken to a processing plant, where a machine separates the seeds from the pods. The seeds are washed, then graded, or sorted into different sizes. Grading is done by passing the seeds over different sizes of wire mesh, allowing only seeds of certain sizes to fall through.

Some seeds undergo special preparation. Split peas, for example, are first soaked and steamed

Harvesting beans by hand on a farm near Faro, Portugal

to soften their outer skin. They are then put into a machine called a splitter, which hurls the peas against metal plates to make them split in half. Then they are graded and polished.

Legumes are usually packaged for sale in stores. Some health food stores, though, keep big bins of legumes, so you can weigh out the amount that you want. Some legumes are prepared and sold in cans ready for immediate use. Canned legumes are convenient but more expensive than dry legumes.

Using a combine harvester to harvest soybeans in Louisiana

Soybeans

Soybeans were first cultivated 5,000 years ago in Eastern Asia. They grow in pods, which can be eaten, but the small, round beans that grow inside the pods are what we usually eat. Whole soybeans need to be soaked for several hours, then cooked for several more before they can be eaten.

Americans get most of their protein from animals in the form of meat, fish, eggs, and milk.

Above: Harvesting soybeans in South Dakota. The United States grows over 60 percent of the world's supply of soybeans.

Opposite: Farming soybeans in Zaire

In poorer countries, people often do not get enough of these foods to eat. Soybeans are an excellent alternative. They are inexpensive, and they contain more protein than any other legume.

The United States grows more soybeans than any other country in the world. Brazil and China are also important producers of soybeans.

Soybeans are made into many products, including flour, oil, margarine, and milk. They are also made into textured vegetable protein (TVP), which tastes and looks like meat. TVP may be made

15

from extruded soy protein or spun soy protein. Extruded soy protein is made by a machine that extrudes, or pushes, soy flour into meatlike shapes. It is usually dried. When you add water, it becomes chewy. Spun soy protein is made by spinning soy protein into fibers. It is sold canned, frozen, and dried. TVP may be mixed with meat or eaten by itself.

More than 95 percent of the soybean meal

Many scientists believe that soybeans could help feed the millions of people who are starving in the world.

These meat-flavored foods are made from a kind of TVP, called textured soya protein.

produced in the United States is used as a protein-rich feed for livestock and poultry. But soybeans grown on an acre of land supplies 10 times more protein than cattle raised for beef on the same land. Many people believe that growing soybeans for human **consumption** could help solve the world hunger problem.

Soybeans also have hundreds of industrial uses. For example, they are used to make paint, glue, paper, plastics, insecticides, and explosives.

Baked beans

In 1895, Heinz was the first manufacturer of baked beans in tomato sauce.

Baked beans are a popular and nutritious food. Americans across the nation eat baked beans at picnics, potluck parties, and for regular lunches and dinners. Baked beans are usually made with red kidney beans or pinto beans and packed in tomato sauce.

Baked beans are made after the harvested dry beans are graded, sorted, and cleaned. The beans are then softened by soaking them in hot water. This is called blanching. Next, the beans are baked and put into cans with the special tomato sauce. The lids are then sealed. The cans pass through a machine that uses steam at a very high temperature to cook the beans. The machine then cools the cans with water, and dries them so they will be ready for labeling.

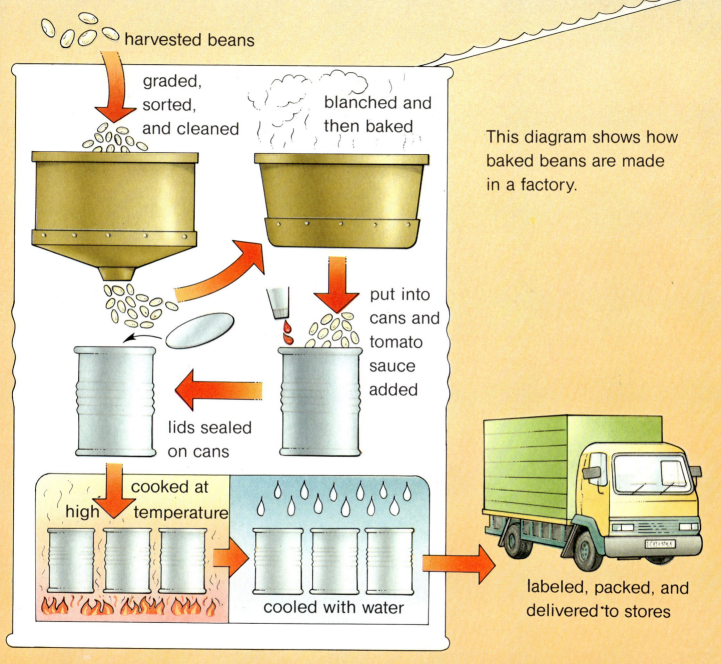

harvested beans

graded, sorted, and cleaned

blanched and then baked

This diagram shows how baked beans are made in a factory.

put into cans and tomato sauce added

lids sealed on cans

cooked at high temperature

cooled with water

labeled, packed, and delivered to stores

Above: Choosing vegetables in a Chinese supermarket

Vegetarianism and veganism

Because they are high in protein, legumes are important in the diet of people who do not eat meat. Such people are called vegetarians or vegans. Vegetarians do not eat meat, poultry, or fish. They do, however, eat dairy products. Vegans do not eat meat or animal products, not even dairy products. Some vegans do not even eat honey because they believe that living creatures should not be used in

Mixed Bean Hotpot

Stuffed Tomatoes

These Buddhist monks are enjoying a vegetarian meal at a temple in Thailand.

Opposite: These tasty vegetarian dishes are prepared without meat. Many people consider a vegetarian diet to be healthier than a meat-eating diet.

any sort of farming at all.

There are many reasons for being a vegetarian. Some religions, such as Buddhism and Hinduism, forbid their followers to eat meat. Some people believe that killing animals for food is wrong, and that modern farming methods cause animals unnecessary suffering. Some people consider a vegetarian diet to be healthier than one containing meat. And others believe that by eating vegetable protein rather than meat protein they are freeing our **natural resources** to feed the world's hungry people.

21

Cooking beans and peas

Lentils, split peas, and canned legumes can be cooked without any prior preparation, but beans and whole peas must be soaked for several hours before cooking. Legumes will absorb twice their own weight in water so use plenty of water for soaking. A quicker method is to cover the legumes with boiling water, let them sit for 10 minutes,

Above: Most legumes should be soaked before they are cooked.

22

Right: Tofu is a soybean curd that is used in many recipes. In this picture, tofu is being prepared in a factory in Tokyo, Japan.

Opposite: A burger made from soybeans and served with vegetables makes a nutritious and satisfying meal.

drain, and pour fresh boiling water over them. This way, they will be ready in an hour.

Recipes will specify the length of time you should cook any particular legume. Lentils cook in about 20 minutes, but larger beans need about an hour. Red kidney beans must be boiled for 15 minutes, drained, then boiled in fresh water. A lot of boiling is needed to destroy a harmful **enzyme** contained in the skin of the bean. Add salt to legumes only after they've been cooked—if you add salt while the beans are cooking, it will make their skins tough.

23

pencil
card
glue
legumes

Things to do with beans and peas

Make a mosaic picture

You will need: a card or strong paper, a pencil or pen, glue, a selection of dried legumes

Draw your design on the card or paper—perhaps a pattern, a face, or a design of your name. Then spread the glue over a little of your design at a time, and fill in your picture with different legumes. For example, little red and yellow split peas are good for hair, larger beans are good for eyes or a nose.

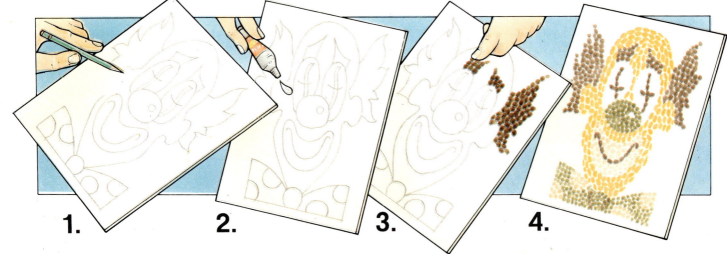

1. **2.** **3.** **4.**

jam jar

mung beans

cheesecloth

rubber band

1.

2.

Grow your own bean sprouts

You will need: mung beans or whole lentils, an empty jam jar, cheesecloth, a rubber band

Place two tablespoons of beans in the jam jar, cover with cold water, and let sit for 8 to 12 hours. Cover the jar with the cheesecloth and fasten with the rubber band. Drain the water through the cheesecloth, refill the jar with fresh water, and drain again. Keep the jar in a warm, dark place, like a cupboard. Rinse and drain twice a day until small **shoots** have grown.

3.

4.

25

Hummus

A creamy dip popular throughout the Middle East, hummus is traditionally served with pita bread, but crusty bread or raw vegetables also go well with it.

You will need, for 4 people:

¾ cup chickpeas (also called garbanzo beans)

4 tablespoons tahini

2 crushed garlic cloves

1 tablespoon olive oil

2 tablespoons lemon juice

salt and freshly ground black pepper

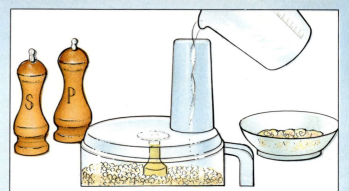

2. Put the chickpeas, tahini, garlic, olive oil, and lemon juice into a blender (be sure to get a grown-up to help you). Blend until smooth, then season with salt and pepper.

1. Soak and cook the chickpeas. Drain and thoroughly rinse them with warm water.

3. Spoon the hummus into a serving dish and pour a little olive oil over the top before serving. Decorate it with lemon slices or olives if you like.

Bean salad

You will need:

¼ cup red kidney beans
1 cup fresh wax beans
1 cup fresh green beans
6 tablespoons olive oil
2 tablespoons wine or cider vinegar
½ teaspoon dry mustard
½ teaspoon sugar
2-3 tablespoons chopped fresh herbs,
  such as parsley and chives
salt and freshly ground black pepper

1. Soak and cook the red kidney beans. Drain and thoroughly rinse the beans with warm water. Rinsing kidney beans is important in getting rid of their harmful enzyme.

2. Cut the wax and green beans into 1-inch lengths, and boil them in a little water for about 3 minutes, until just tender.

3. Put the oil, vinegar, mustard, and sugar into a clean screw-top jar and shake until blended.

4. Mix the beans together in a bowl and pour on the dressing. Doing this when the beans are still warm will help the flavors blend. Add the herbs, salt, and pepper to taste, and mix gently. Serve the salad chilled.

Cheesy lentil bake

You will need, for 4 people:

1 medium-sized onion
2 tomatoes
1¼ cup split lentils
2 cups water
1 cup grated cheddar cheese
½ teaspoon dried oregano or
    marjoram
salt and freshly ground black pepper
a little butter or margarine

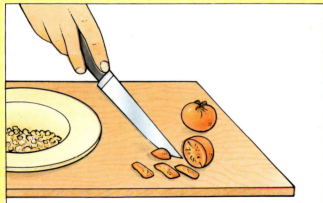

1. Chop the onion and slice the tomatoes (get a grown-up to help you).

2. Put the chopped onion and lentils into a saucepan with the water, bring to a boil, and simmer gently for about 20 minutes, until the water is absorbed and the lentils are tender.

28

3. Mix in half the cheese and the oregano or marjoram, and season with salt and pepper.

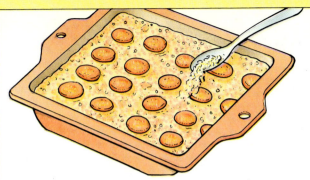

5. Arrange the tomato slices over the top and sprinkle with the rest of the grated cheese. Dot the top with a little butter or margarine. Bake in an oven heated to 400°F for 25-30 minutes, until the top is golden and crunchy.

4. Press the mixture into a greased loaf pan.

6. Slice the cheesy lentil bake, and serve with salad or potatoes and vegetables.

Glossary

amino acids: the building blocks of protein. Some are made in our bodies, but others can only be found in the foods we eat.

consumption: the act of eating

cultivating: growing and caring for plants; farming

enzyme: a substance that makes specific chemical reactions happen within the body

exporters: businesses or a country that sells products to foreign countries

natural resources: materials, found in nature, that people use

nutritious: healthy food that promotes growth

protein: combination of amino acids that help the body grow and mend itself.

shoots: the first growths from seeds

Photo acknowledgments

The photographs in this book were provided by: pp. 6, 11, 12, 22, J. Allan Cash; pp. 8, 18, H.J. Heinz Company; pp. 10, 13, Holt Studios; pp. 9, 14, 15, 16, 21, The Hutchison Library; pp. 7, 20, Christine Osborne; p. 17, Wayland Picture Library; p. 23, Zefa.

Index